I study all aspects
of the ocean.

Oceanographer
[oh-shun-og-ruh-fer]

I'm in charge of
all electrical systems

Electrician

I'm in charge of the safety
and daily operations of
the research vessel and crew.

Research Vessel Captain

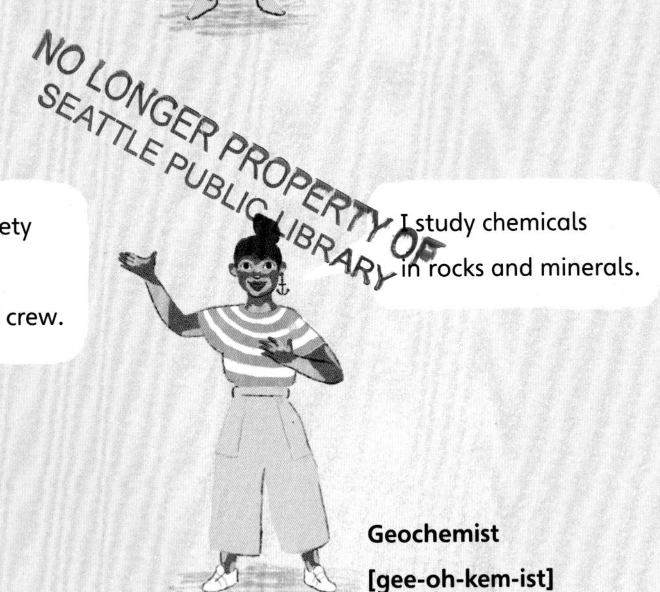

I study chemicals
in rocks and minerals.

Geochemist
[gee-oh-kem-ist]

I study human history
through remains
and artifacts
found underwater
or along the shore.

Marine Archaeologist
[ma-reen ar-kee-all-uh-jist]

I pilot (drive)
and repair submersibles.

Submersible Pilot
[sub-mer-suh-bul py-let]

To Will, for everything.

To Ann Schwab and the Children's Library team,

for creating a brave space to make mistakes and grow. —ASF

To Uncle Hugh and Auntie Nina;

thanks for always being a safe harbor. —ACM

Library of Congress Cataloging-in-Publication Data

Names: Forrester, Amy Seto, author. | Musser, Andy Chou, illustrator.
Title: Search for a giant squid : pick your path / by Amy Seto Forrester and Andy Chou Musser.
Description: San Francisco : Chronicle Books, [2023] | Series: "Science Explorers" ; book 1 | Cover title. |
Includes bibliographical references.
Identifiers: LCCN 2022022130 | ISBN 9781797213934 (hardcover)
Subjects: LCSH: Giant squids—Habitat—Juvenile literature. | Giant squids—Research—Juvenile literature.
Classification: LCC QL430.3.A73 F67 2023 | DDC 594/.58—dc23/eng/20220603
LC record available at https://lccn.loc.gov/2022022130

Manufactured in China.

Design by Sara Gillingham Studio.
Typeset in Heinemann.

10 9 8 7 6 5 4 3 2 1

Chronicle Books LLC
680 Second Street
San Francisco, California 94107

SCIENCE EXPLORERS

Search for a GIANT SQUID

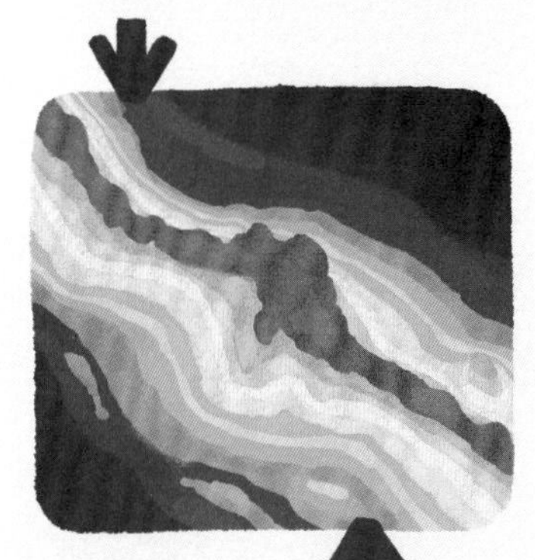

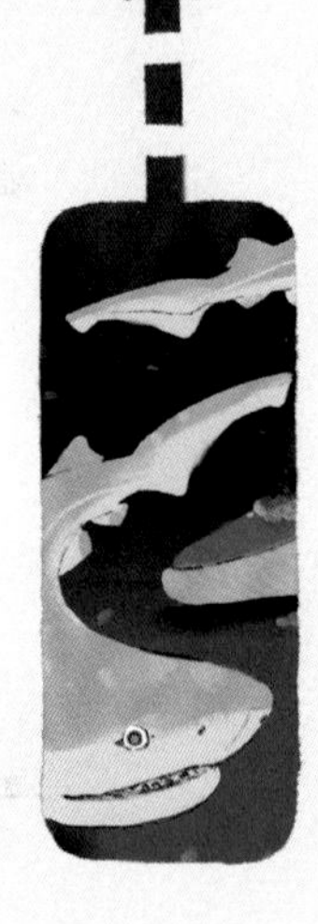

Amy Seto Forrester &
Andy Chou Musser

chronicle books · san francisco

Join the Team

We need your help making choices on our expedition.
Some choices will lead us closer to a giant squid; other choices might not.
But we'll still see and learn amazing things!
See a word you don't know?
Check the glossary on page 90.

Giants of the Deep

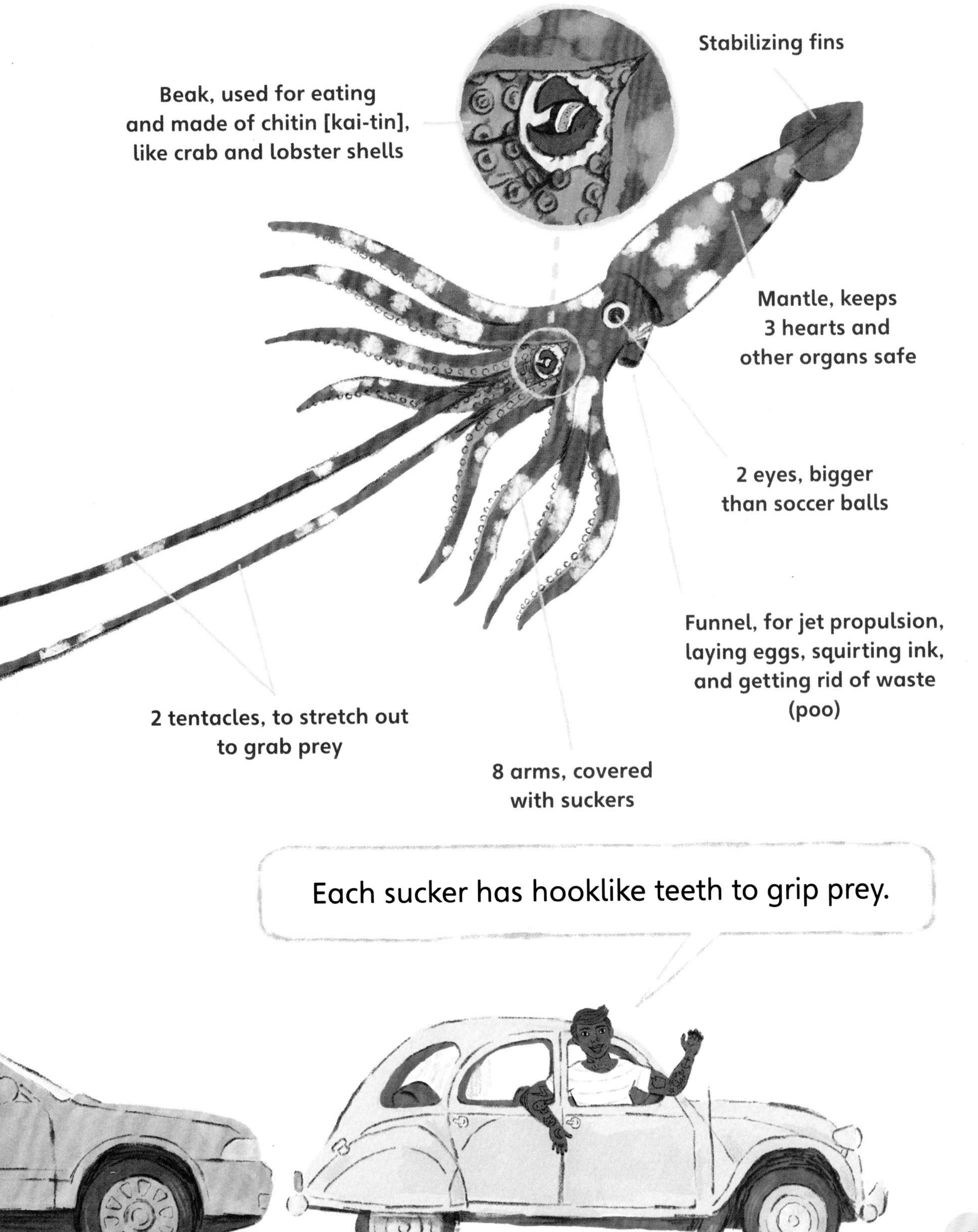
Stabilizing fins
Beak, used for eating and made of chitin [kai-tin], like crab and lobster shells
Mantle, keeps 3 hearts and other organs safe
2 eyes, bigger than soccer balls
Funnel, for jet propulsion, laying eggs, squirting ink, and getting rid of waste (poo)
2 tentacles, to stretch out to grab prey
8 arms, covered with suckers
Each sucker has hooklike teeth to grip prey.

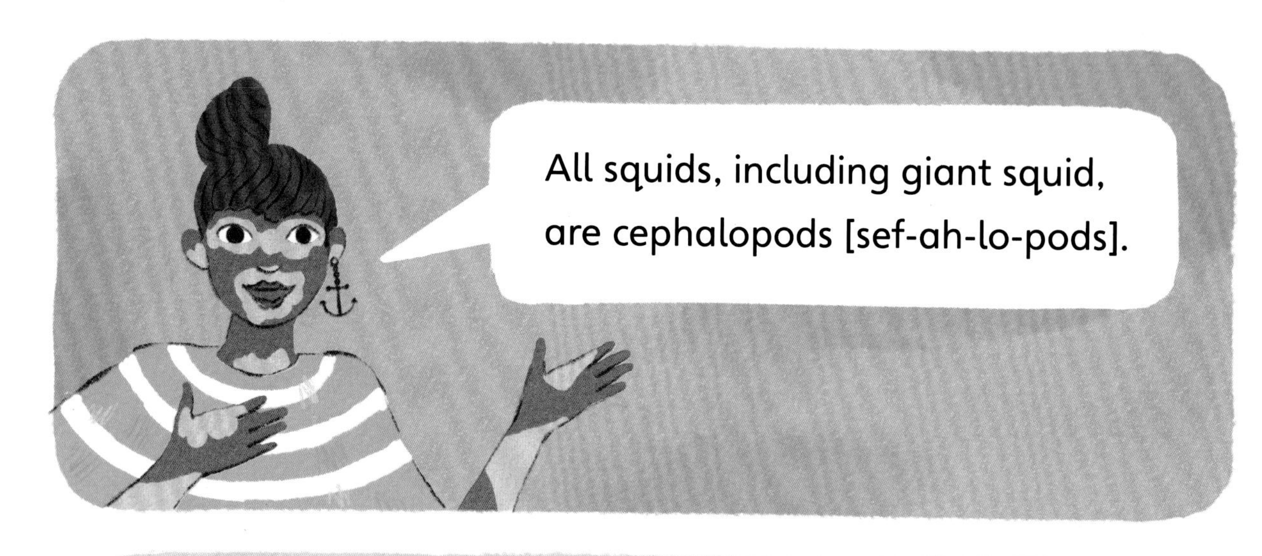
All squids, including giant squid, are cephalopods [sef-ah-lo-pods].

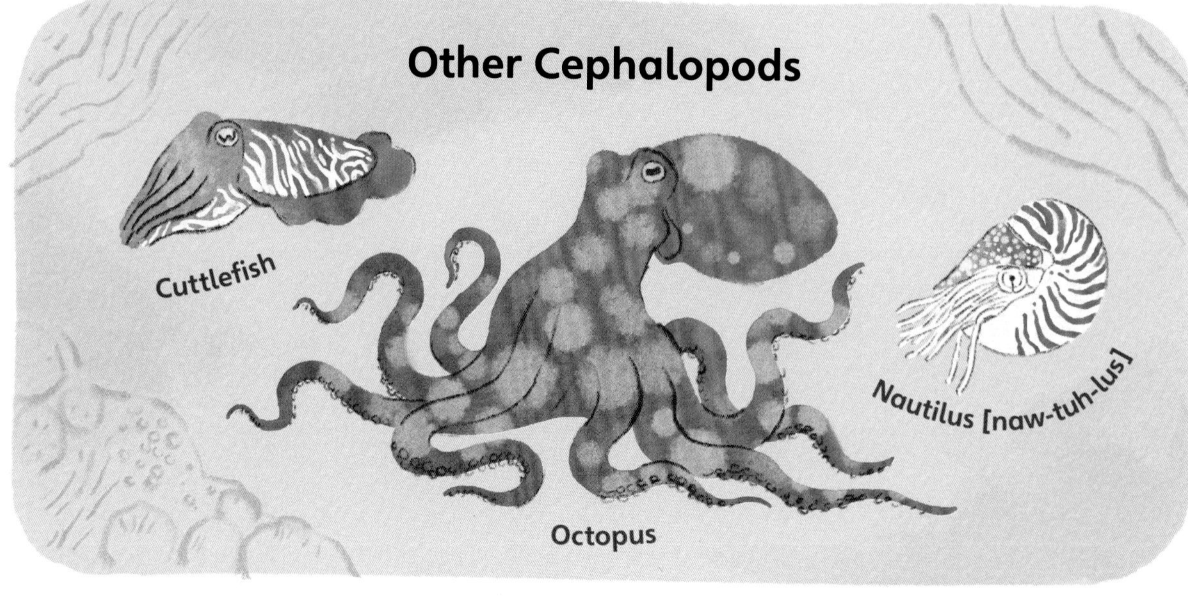
Other Cephalopods
Cuttlefish
Octopus
Nautilus [naw-tuh-lus]

Scientists who study cephalopods are called *teuthologists* [tooth-all-uh-jists].

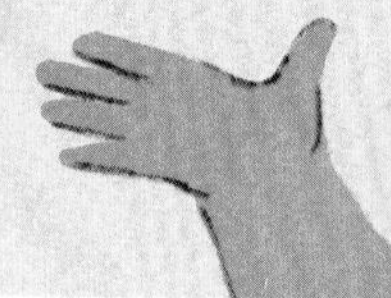

Even though teuthologists have learned a lot about giant squid, they still have many questions.

How many giant squid are there?

How many babies do giant squid have?

Are giant squid hurt by ocean pollution?

We don't know!

Why don't teuthologists know more?

For a long time, scientists couldn't dive deep enough or stay underwater long enough to study giant squid in the ocean.

Ocean Zones

Sunlight Zone: 0–660 feet (0–200 metres)

Twilight Zone: 660–3,300 feet (200–1,000 metres)

Midnight Zone: 3,300–13,000 feet (1,000–4,000 metres)

Abyssal Zone: 13,000–20,000 feet (4,000–6,000 metres)

Hadal Zone: 20,000–36,000 feet (6,000–11,000 metres)

The giant squid's natural habitat is the twilight zone.

If you tried to swim there, your body would get squished by water pressure.

Scientists need underwater vehicles to travel safely to the twilight zone.

We will use an underwater vehicle called a *submersible* [sub-mer-suh-bul].

Submersibles are small and easy to turn.

Submarines are underwater vehicles too.

But they are too big and slow for this kind of expedition.

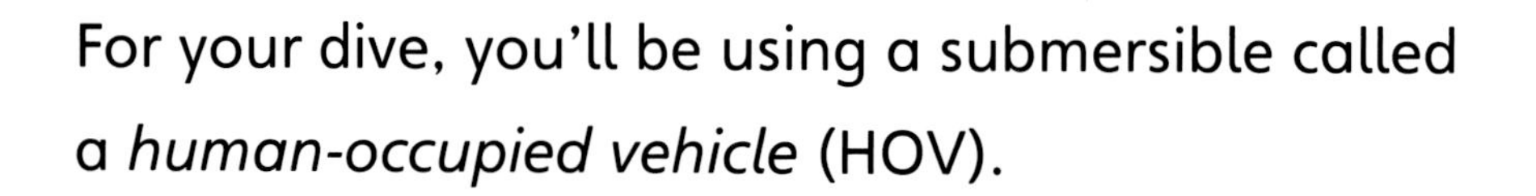

For your dive, you'll be using a submersible called a *human-occupied vehicle* (HOV).

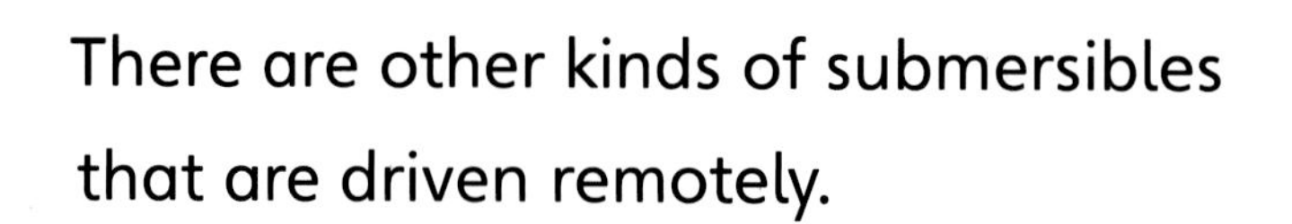

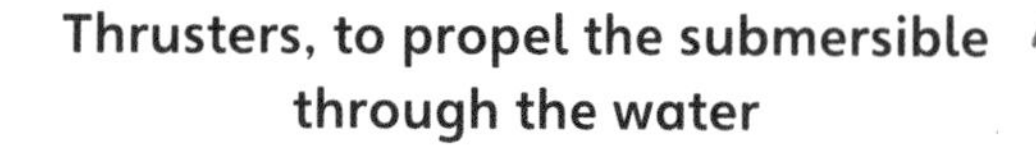

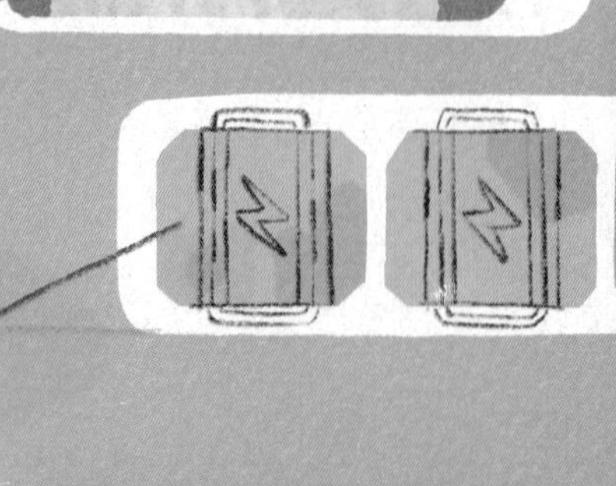

To get to your dive site, your submersible will be transported by a ship called a *research vessel.*

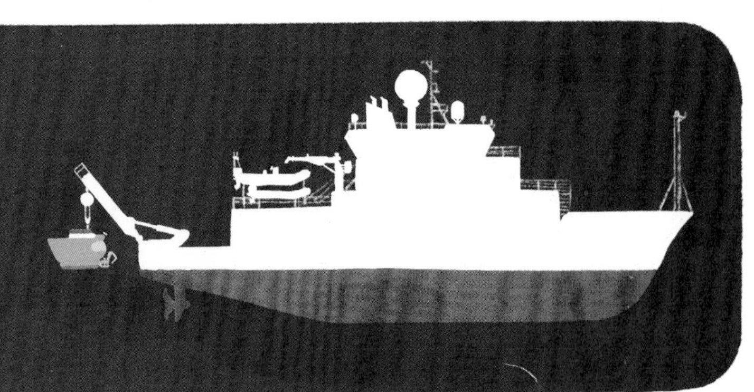

Getting Ready

Before your dive, you and your team work together to make a science plan.

This plan lists everything you want to do on your dive.

Making a science plan takes a lot of time, but it's also a lot of fun!

I would love to get rock samples from that area too!

Can you add that to our science plan?

You got it!

Let's Get Packing!

It's important to pack everything you will need on your expedition.

There are no shops at sea!

The twilight zone is cold!

Bring warm clothes.

Cotton or wool is safest in case of a fire.

Expeditions are exciting but hard.

Bring things that remind you of home.

Be ready to fix anything that breaks.
Duct tape is always useful!
Don’t forget the zip ties!

Preparing the Research Vessel

Bridge
Look at the weather forecast with the captain.

Work Boats
A small team will be on the water to help during your dive.

A-Frame
This machine lifts your submersible into and out of the water.

Hangar
Your submersible is stored here until your dive.

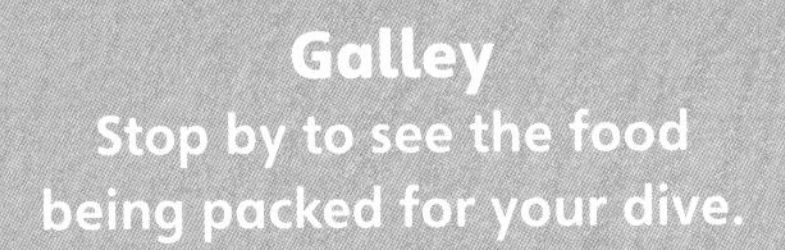

Galley
Stop by to see the food
being packed for your dive.

My favorite meal is
a peanut butter and honey sandwich,
an apple, and a candy bar!

Head
Be sure to use the bathroom
before your dive.

In a submersible,
the only way to pee
is into a bottle.

Labs
Get your equipment ready for samples
you'll bring back from your dive.

Berths
Everyone has a place
to sleep.

Your Adventure Begins

Now it's time for YOU to make some choices so your adventure can begin

Start by picking a pilot.

NAME: Robin

To pick Robin, go to page 79.

I learned to build and fix electrical equipment by working with other pilots and engineers.

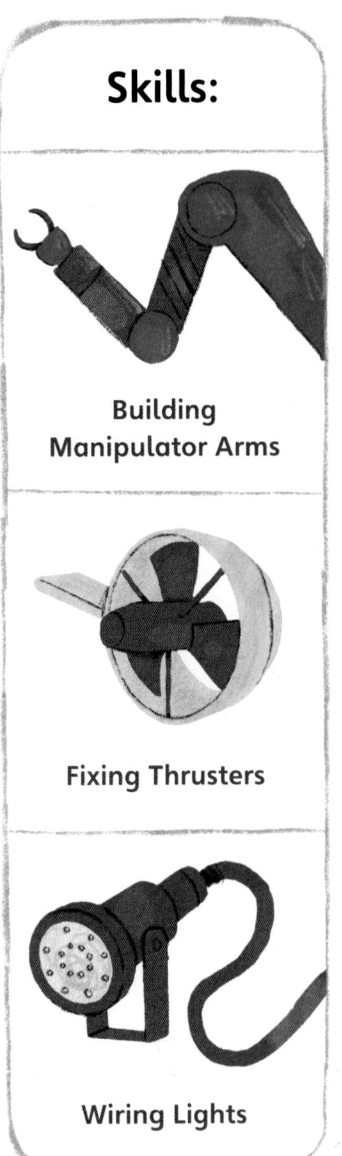

Skills:

Building Manipulator Arms

Fixing Thrusters

Wiring Lights

NAME: Scout

To pick Scout, go to page 61.

Pick a Submersible

GALAXY

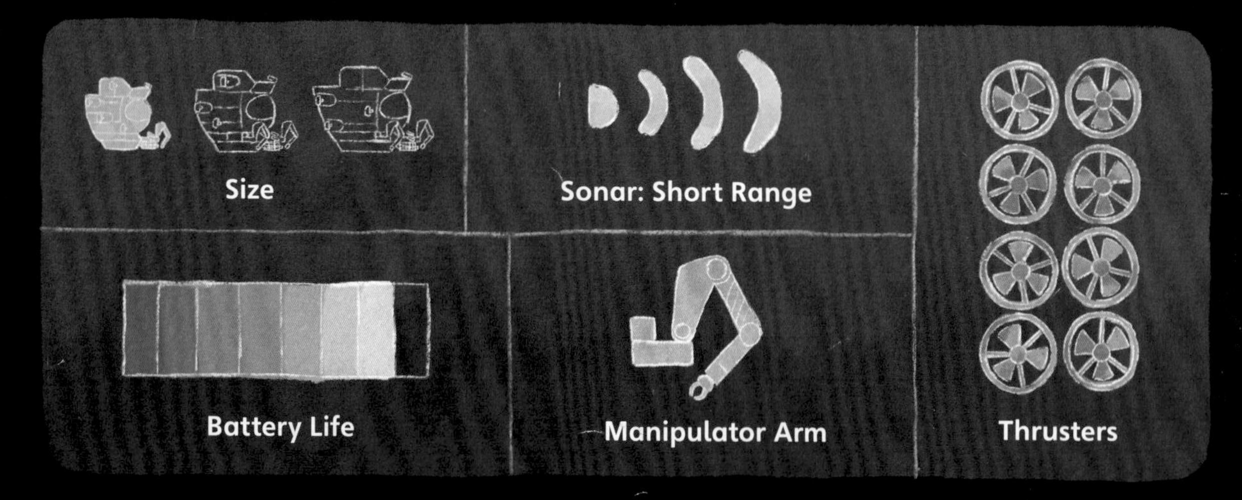

If Scout is your pilot, go to page 75 to pick *Galaxy*.

If Robin is your pilot, go to page 42 to pick *Galaxy*.

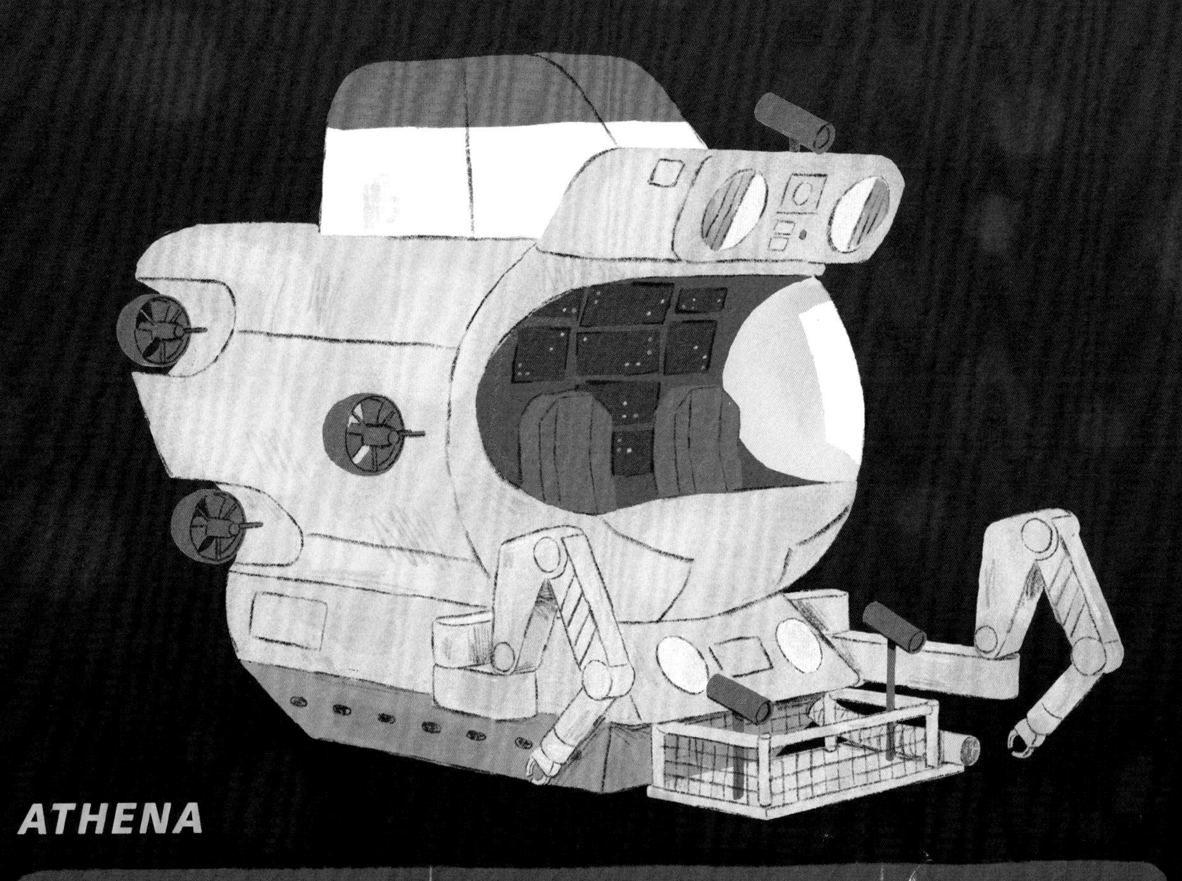

ATHENA

Size

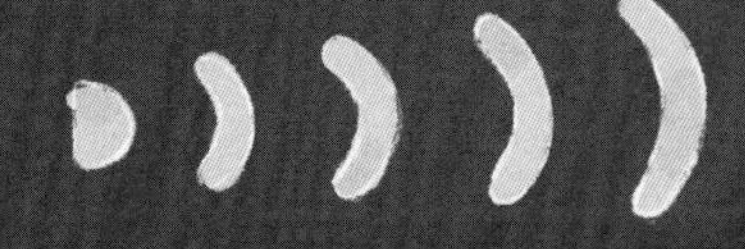

Sonar: Long Range

Thrusters

Battery Life

Manipulator Arms

If Scout is your pilot, go to page 31 to pick *Athena*.

If Robin is your pilot, go to page 57 to pick *Athena*.

Go to Your Dive Site!

South Atlantic Ocean

Dead giant squid have washed up on beaches in South Africa.

Could giant squid live nearby?

To pick this dive site, go to page 28.

Giant squid are found in almost all the oceans of the world.

Gulf of Mexico

A giant squid was seen by scientists in the waters between the United States, Mexico, and Cuba.

Maybe there's a chance you could see more!

To pick this dive site, go to page 54.

Pacific Ocean

The first video of a live giant squid was recorded near Japan.

Maybe giant squid like to eat fish that live around seamounts in this part of the ocean.

To pick this dive site, go to page 62.

South Atlantic Ocean

The South Atlantic Ocean is part of the Atlantic Ocean, the second largest ocean in the world.

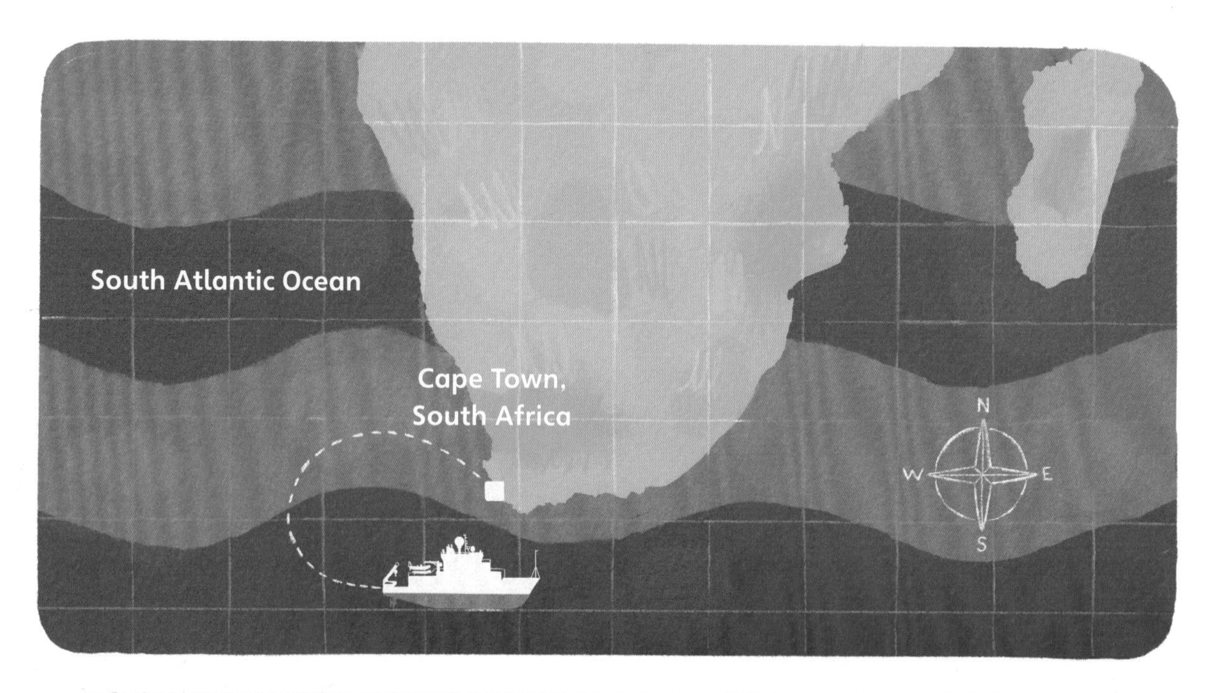

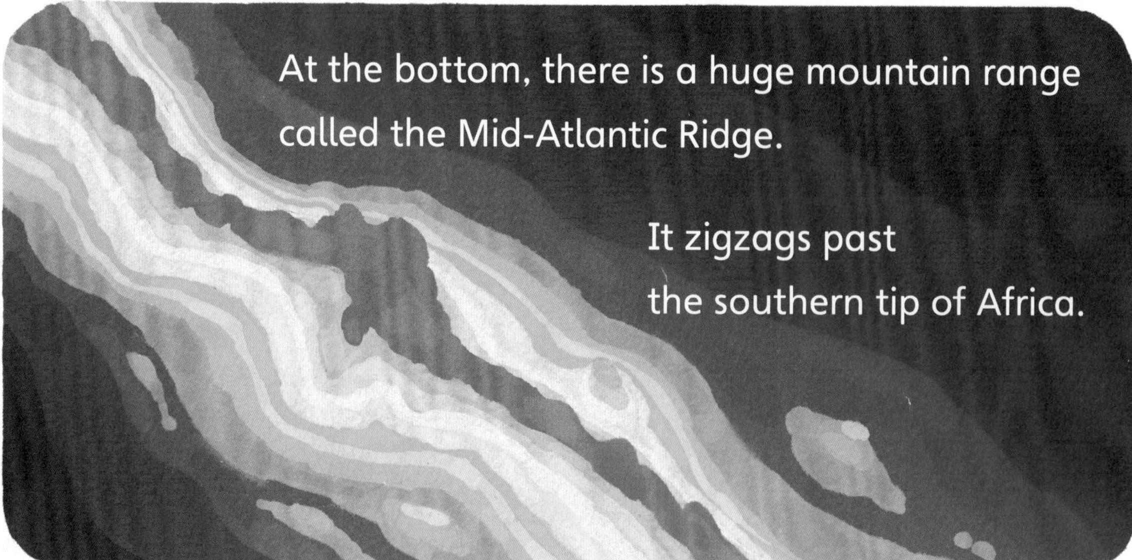

But you can't see any of those mountains from the research vessel.

What other hidden things will you see under the ocean?

Yellow-nosed albatross

Orcas

Your adventure continues on page 56.

Your pilot points the thrusters down.

You quickly rise above the tidal current.

You're safe!

Look, there's a wonderful firefly squid!

They can change color by shrinking
or expanding pigment cells in their skin
called *chromatophores* [cro-ma-toe-fors].

BEEP!
BEEP!

Someone is calling you on the radio.

Your adventure continues on page 41.

Athena Is Your Submersible!

It's a big submersible
with lots of room for equipment!

To pick your dive site, go to page 26.

It's a huge blob of squid eggs!

Maybe they are giant squid eggs?

Scientists don't know
what giant squid eggs look like
or what baby giant squid eat.

Your discovery could change that!

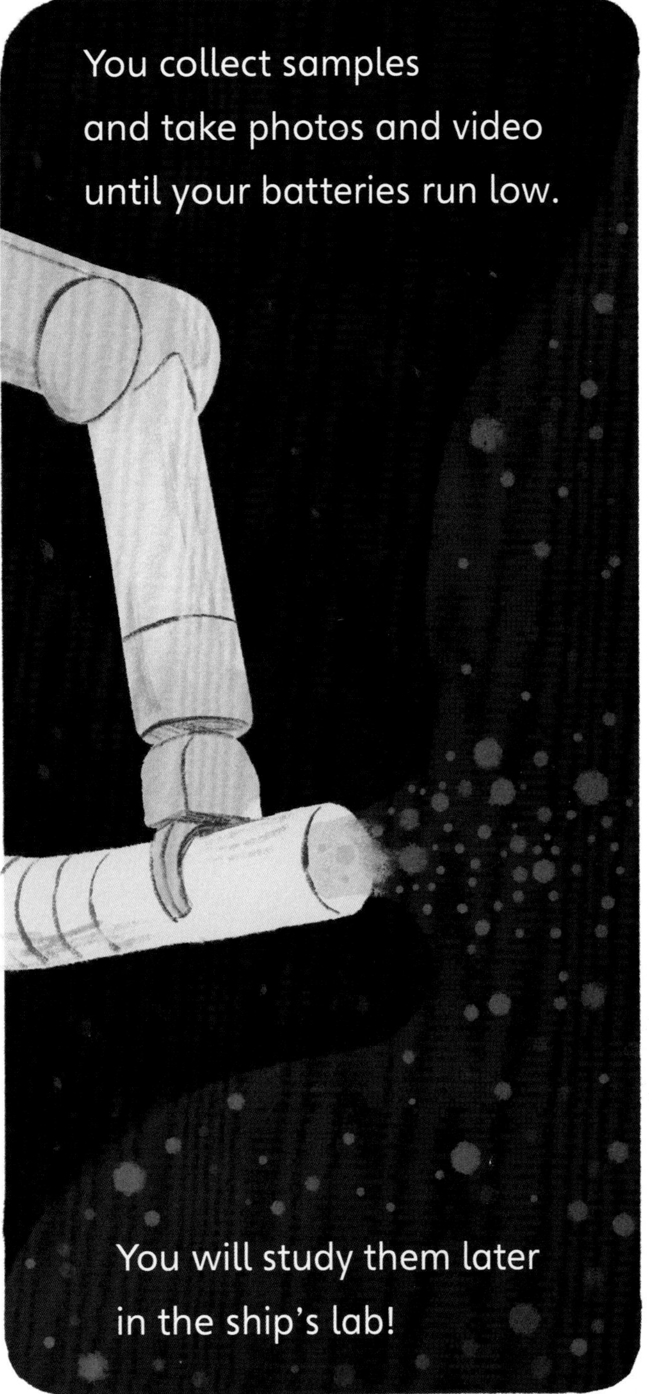

You collect samples
and take photos and video
until your batteries run low.

You will study them later
in the ship's lab!

End of dive.

To try again, go back to page 22.

Your pilot navigates to the left.

Look, it's a giant siphonophore [si-faa-nuh-for]!

These compound animals are made of
hundreds of smaller animals called *zooids* [zoe-ids].

Giant siphonophores are fragile.

They can't be taken back to the lab.

So, you take video and photos
until your submersible's batteries run out.

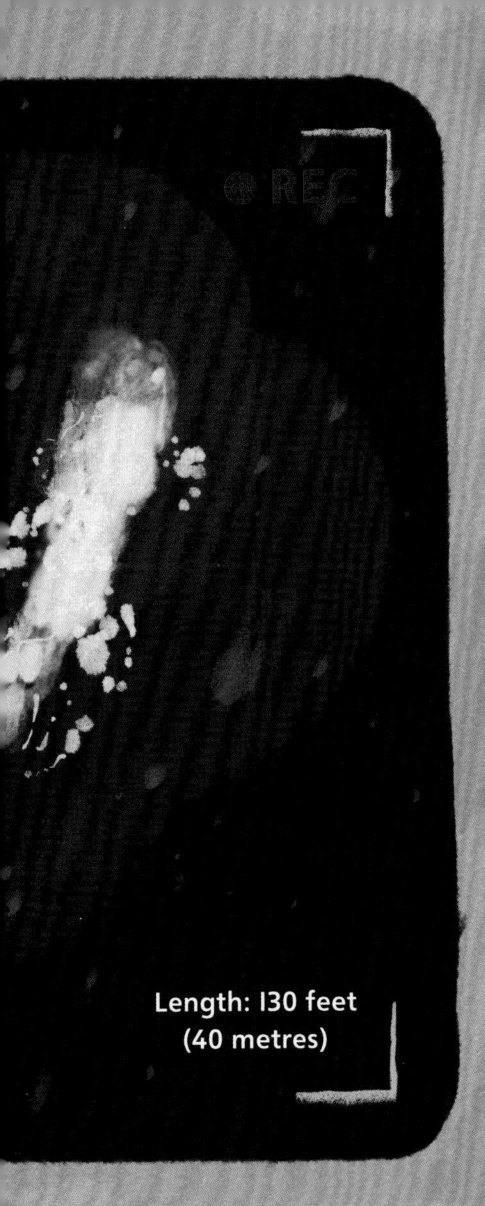

Length: 130 feet
(40 metres)

Zooids work together as a team.

Each one has a special job:
eating, swimming,
making light with bioluminescence
[bye-oh-loom-in-eh-cents],
and more.

You didn't find a giant squid.

Instead, you found this dazzling giant siphonophore!

End of dive.

To try again, go back to page 22.

Your submersible goes deeper.

There's a shipping container on the seafloor!

It must have fallen off a cargo ship.

You can't see what's inside

So, you take samples of the mud around the container to study later for clues.

Suddenly, something big swims past your camera.

If you're in *Galaxy*, go to page 67.
If you're in *Athena*, go to page 74.

Running into one of those could damage the submersible.

Your adventure continues on page 43.

You're being hit by a tidal current!

The force could push your submersible into rocks and damage it.

Quick, should your pilot point the thrusters up or down to get out of the tidal current?

To point the thrusters up, go to page 44.

To point the thrusters down, go to page 30.

Your adventure continues on page 66.

Storms above the water can create underwater currents even stronger than tidal currents.

You didn't find a giant squid,
but you did get photos and video of a wonderful firefly squid!

End of dive.

To try again, go back to page 22.

Galaxy Is Your Submersible!

To pick your dive site, go to page 26.

Your sonar may be broken, but there are still exciting things to see!

There's a red paper lantern jellyfish!

When surprised, they crumple their mantles.

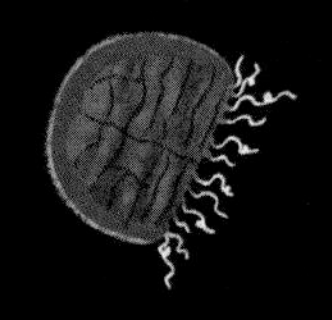

Scientists are still learning why they do this.

WARNING!

Suddenly, your submersible stops.

Your adventure continues on page 76.

Your pilot points
the thrusters up.

SLAM!

You hit the ocean floor.

Oh no! Your submersible is stuck in the mud!

End of dive.

To try again, go back to page 22.

The barreleye glides through marine snow.

Marine snow is made of pieces of dead plants and animals, sand, poo, and more!

ZAP!

Your submersible's computer crashes.

Your sonar stops working!

Without sonar, you can't see anything in the dark water until you are right next to it.

Can your pilot fix the sonar?

If Scout is your pilot, go to page 38.

If Robin is your pilot, go to page 81.

Lights fixed!

Now we can use our cameras again.

Look, there's a deep-sea dragonfish!

They light up the tips of their chin barbels using bioluminescence [bye-oh-loom-in-eh-cents].

Wait, what's that?!

Your adventure continues on page 32.

It's the tail fin of a bluntnose sixgill shark!

Your pilot follows it to a group of bluntnose sixgill sharks, known as a *shiver*.

They are feeding on a fin whale.

When a whale dies, it sinks into the ocean and becomes a whale fall.

A whale fall is food for many animals.

The sharks turn toward your submersible!

Do they think you want their food?

If you're in *Galaxy*, go to page 80.

If you're in *Athena*, go to page 40.

You use all your battery power exploring.

End of dive.

To try again, go back to page 22.

There it is again!

Your submersible's lights go out.

The electrical wiring must have gotten loose when the sharks bumped the submersible.

Can your pilot fix the lights?

If Scout is your pilot, go to page 47.
If Robin is your pilot, go to page 72.

Look!

Two tentacular clubs
and eight arms.

Could it be a giant squid?

Your adventure continues on page 84.

Gulf of Mexico

The Gulf of Mexico is the world's largest gulf.

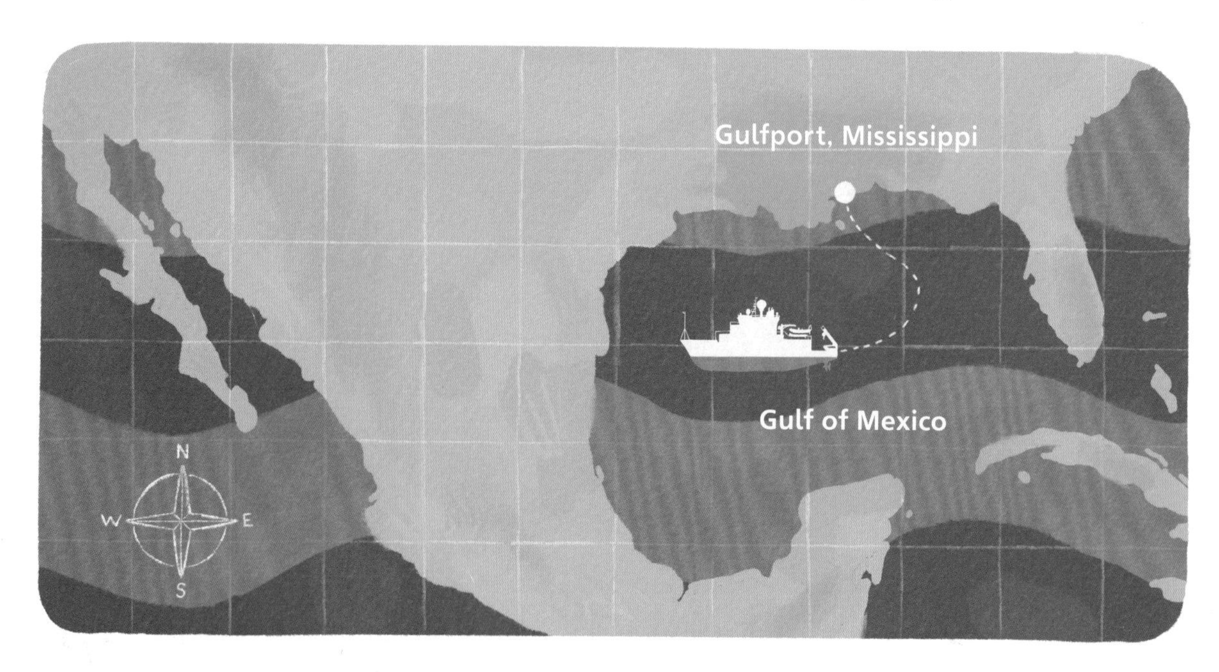

A gulf is a large part of an ocean surrounded by land, except for a narrow opening.

An hour ago, you were standing on the ship's deck in the hot sun.

Now you are on your way to the cold, dark twilight zone.

As your submersible dives deeper, you see plastic everywhere.

Humans can also get sick from eating these animals and from drinking water filled with microplastics.

Your adventure continues on page 48.

Athena Is Your Submersible!

To pick your dive site, go to page 26.

The goblin shark leads you to a shipwreck!

Shipwrecks create unique ecosystems with many animals and plants living on or around them.

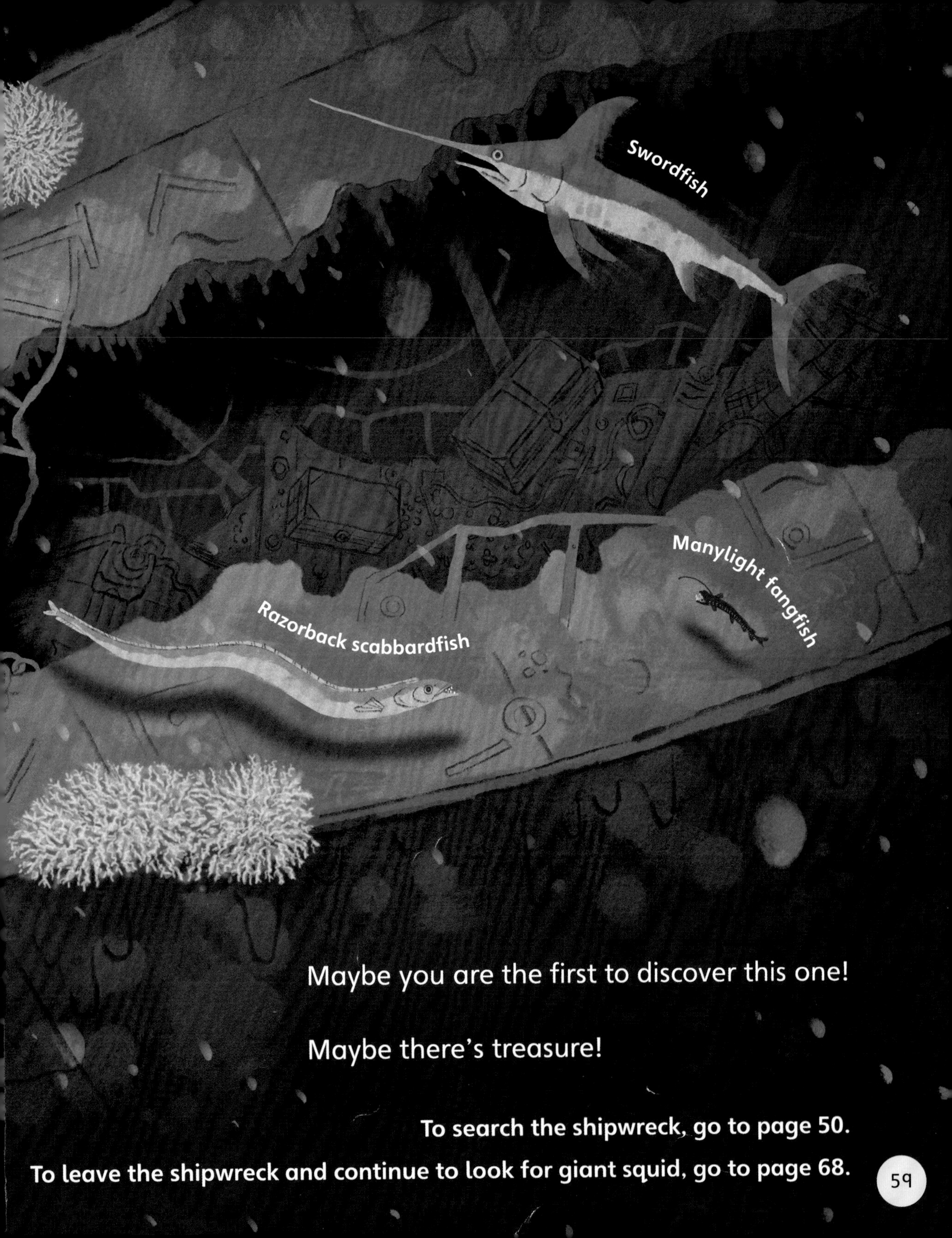

Maybe you are the first to discover this one!

Maybe there's treasure!

To search the shipwreck, go to page 50.

To leave the shipwreck and continue to look for giant squid, go to page 68.

Two black-eyed squid!

They are famous for carrying their eggs.

This is called *brooding*.

I can't wait for scientists to see this video!

End of dive.

To try again, go back to page 22.

Scout Is Your Pilot!

I'm excited to dive with your team!

Are you ready to pick a submersible?

Look for one with more thrusters for more power.

To pick your submersible, go to page 24.

Pacific Ocean

The Pacific Ocean is the largest, deepest ocean in the world!

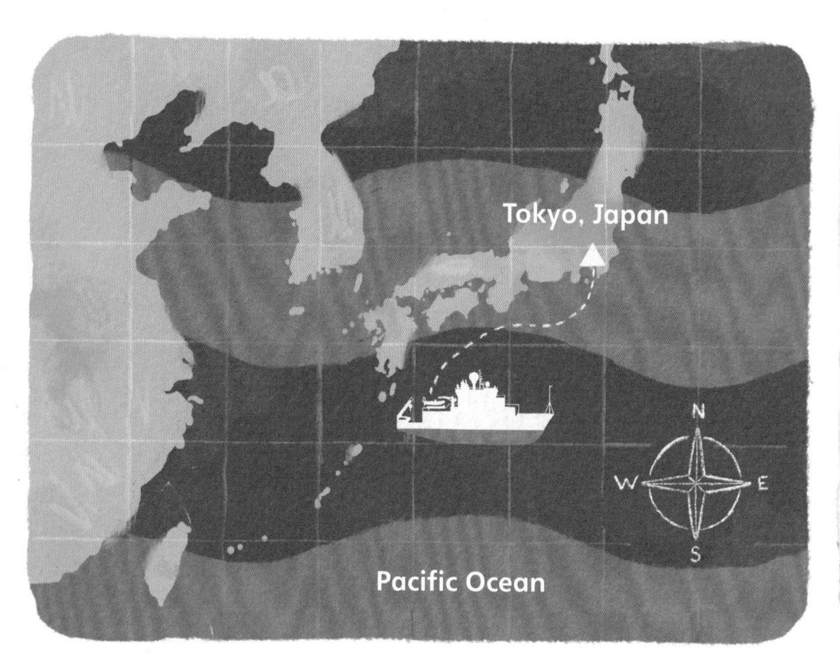

The name Pacific means *peaceful.*

Luckily, today the Pacific is as calm as its name.

Your adventure continues on page 70.

Your submersible glides through the sunlight into the twilight zone.

In less than 15 minutes, the water changes from bright blue to dark blue to black.

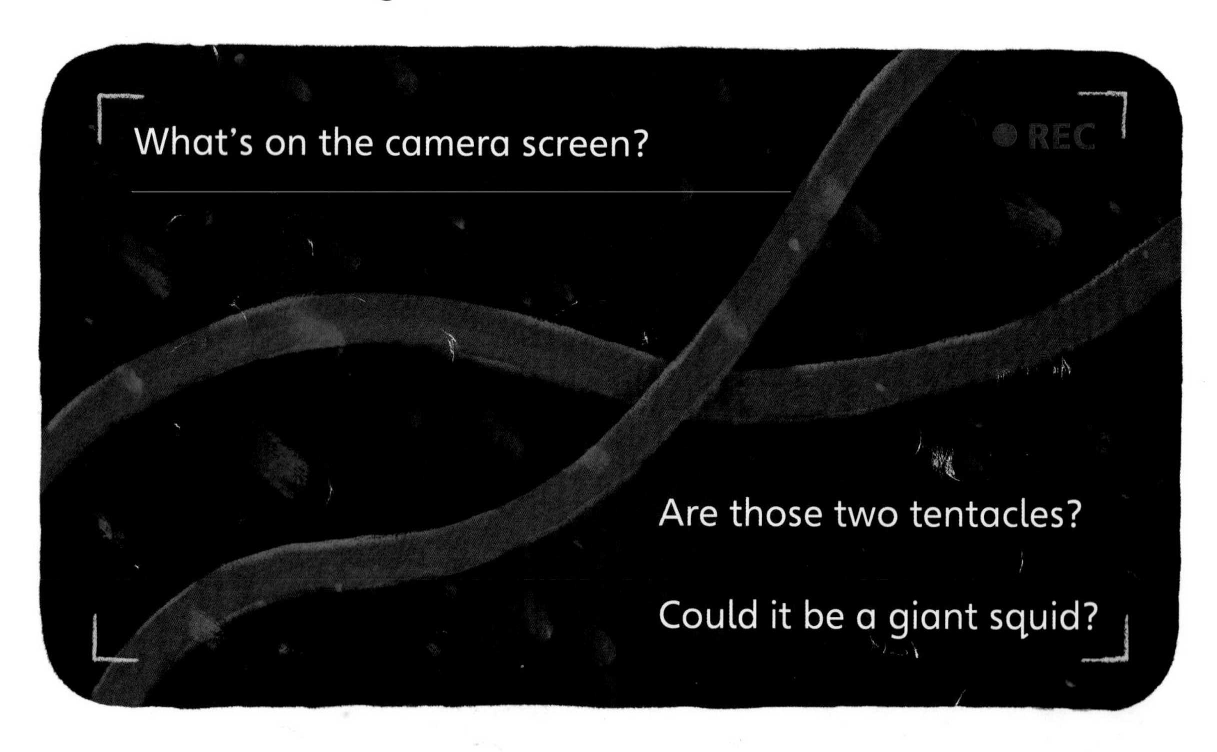

To follow the animal, go to page 78.

To continue deeper into the twilight zone, go to page 36.

Your pilot navigates to the right.

Slowly, the water lights up with more and more bioluminescence [bye-oh-loom-in-eh-cents].

There must be more oxygen here if there are this many animals.

You made it out of the OMZ!

Suddenly, a big animal swims by the camera.

Is that a stabilizing fin?

Your adventure continues on page 52.

There's a fishing net stuck in one of *Athena*'s thrusters.

You need to go back to the ship to fix the thruster.

End of dive.

To try again, go back to page 22.

It's a group of oceanic manta rays!

They have the biggest brains of any fish scientists have found so far.

They're feeding on fish and tiny animals like pink shrimp and krill.

Is that a lake up ahead?

REC

Wait, how can a lake be underwater?

Your adventure continues on page 82.

As you leave the shipwreck, animals swim by.

These fish are also called *owlfish* because of their huge eyes.

These animals can breathe in areas with very little oxygen in the water.

This could be an oxygen minimum zone (OMZ)!

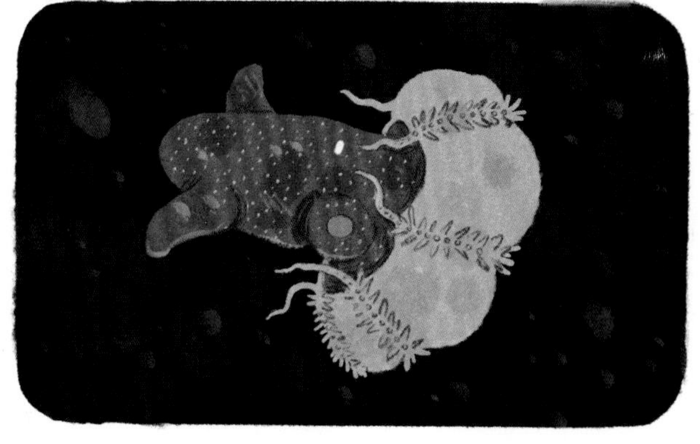

Vampire squid protect themselves by folding up their webbed arms.

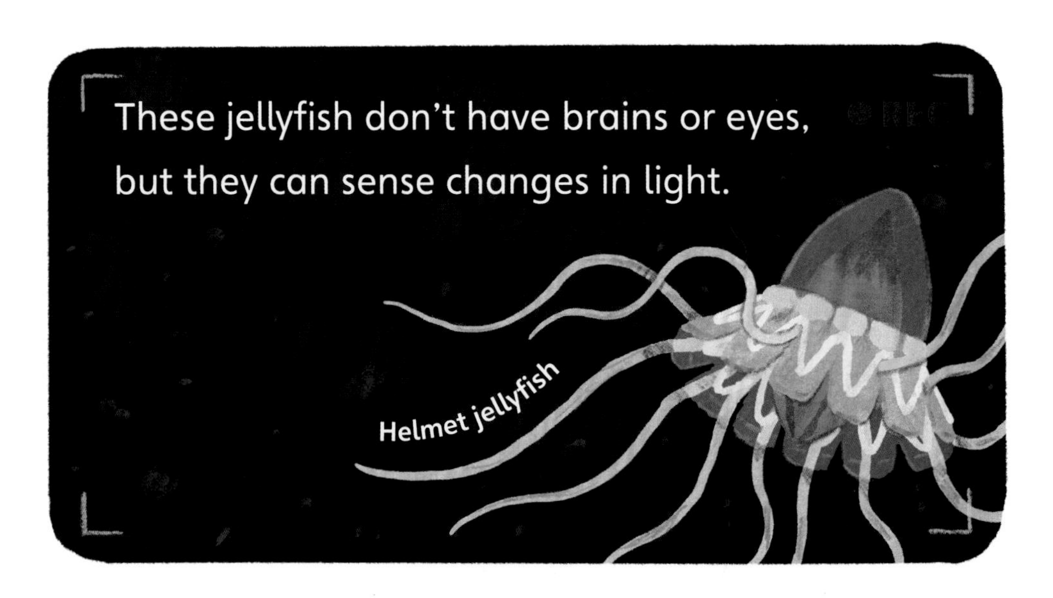

The water in an OMZ is filled with sulfur [sul-fer] made by microbes.

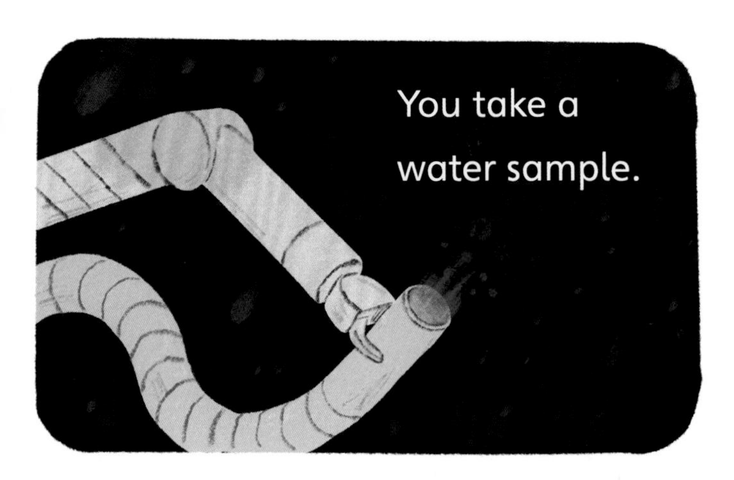

Scientists don't know if giant squid can breathe in OMZs.

Let's find an area of the ocean with more oxygen.

Which way should we go?

To turn left, go to page 34.

To turn right, go to page 65.

Your submersible reaches the twilight zone quickly!

There are more fish in this zone
than in all the other ocean zones combined!

Barreleyes can rotate
their eyes to look up through
their see-through heads.

Goblin sharks can stretch out
their jaws to catch prey!

To follow the barreleye, go to page 46.
To follow the goblin shark, go to page 58.

You can't follow the sunfish today.

It's time to go back to the ship.

The crew is excited to study your photos, video, and samples!

Wow!

End of dive.

To try again, go back to page 22.

The lights flash once.

They flash twice!

Then they go dark.

I can't fix the lights.

No more exploring today.

Suddenly, lights flicker in the water.

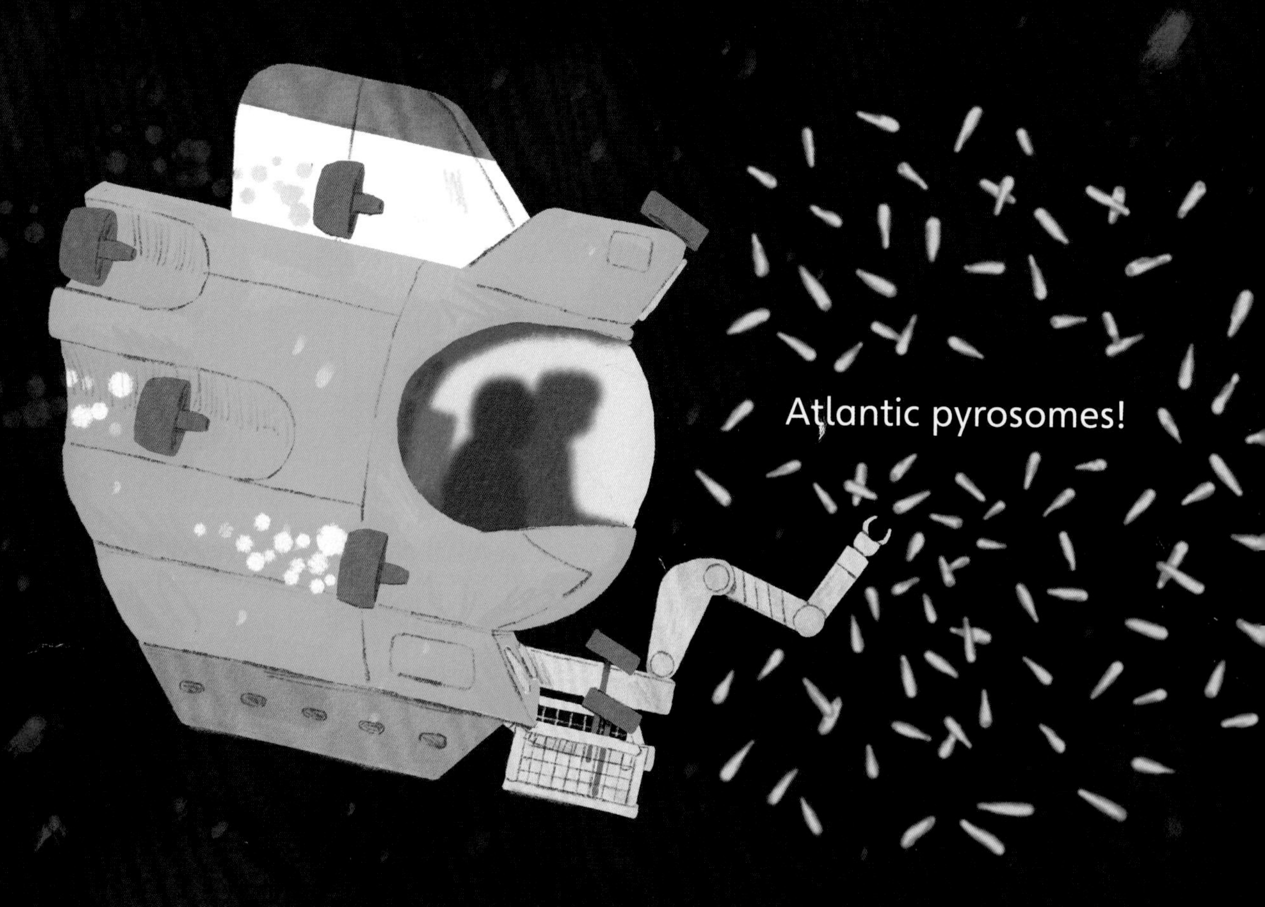

Atlantic pyrosomes make their own light using bioluminescence

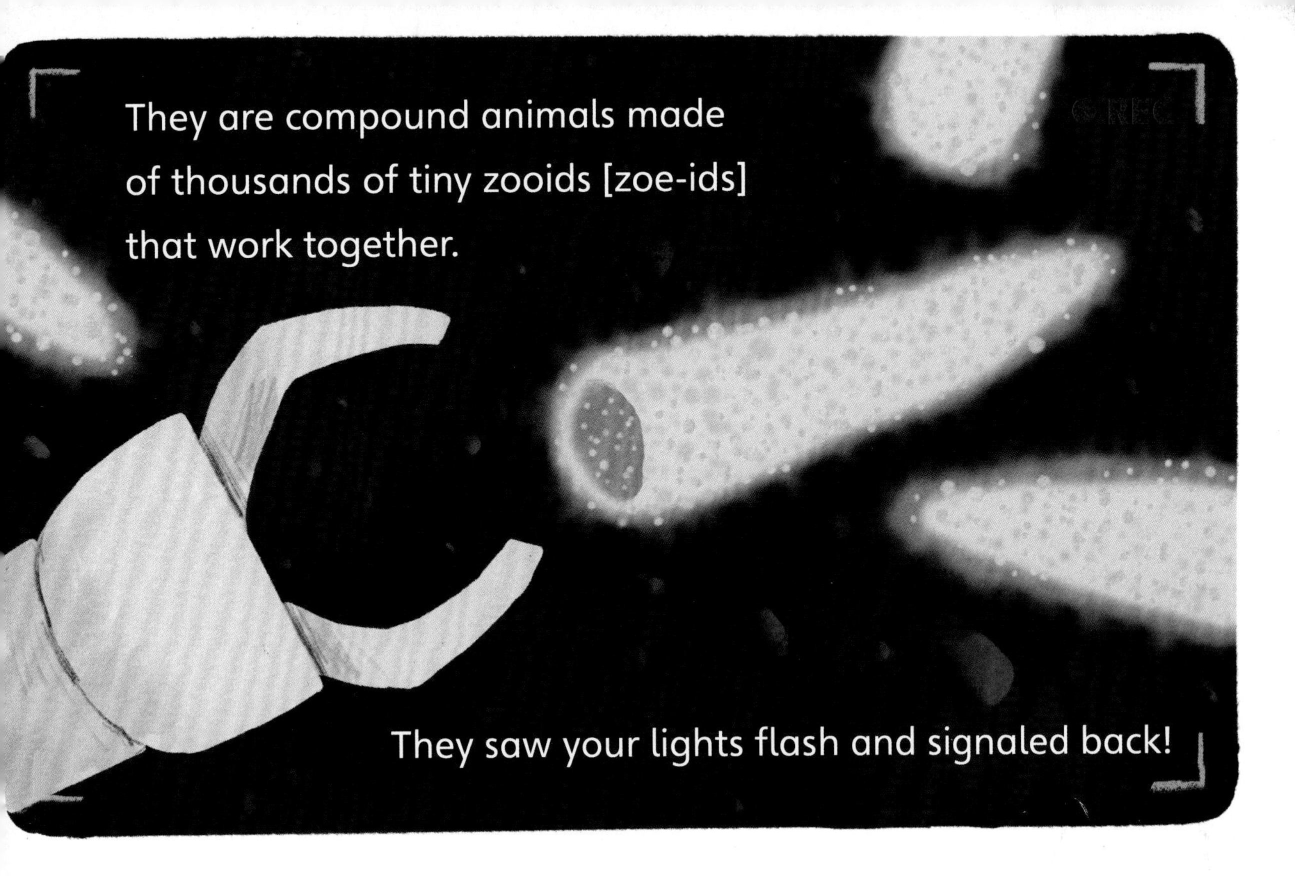

Scientists are still learning what this signal means.

We don't know if giant squid use bioluminescence, but we do know that they're attracted to it.

Now it's time to return to the ship, but maybe these photos and video will help scientists learn more!

End of dive.

To try again, go back to page 22.

Athena speeds forward, but the big animal is too fast.

Then you see a smaller animal.

Your adventure continues on page 71.

Galaxy Is Your Submersible!

To pick your dive site, go to page 26.

WHAM!
Your back thruster smashes
against a seamount.

Our sonar is broken,
so we didn't see the seamount.
You must go back to the ship
to fix your thruster.
Error!

End of dive.

To try again, go back to page 22.

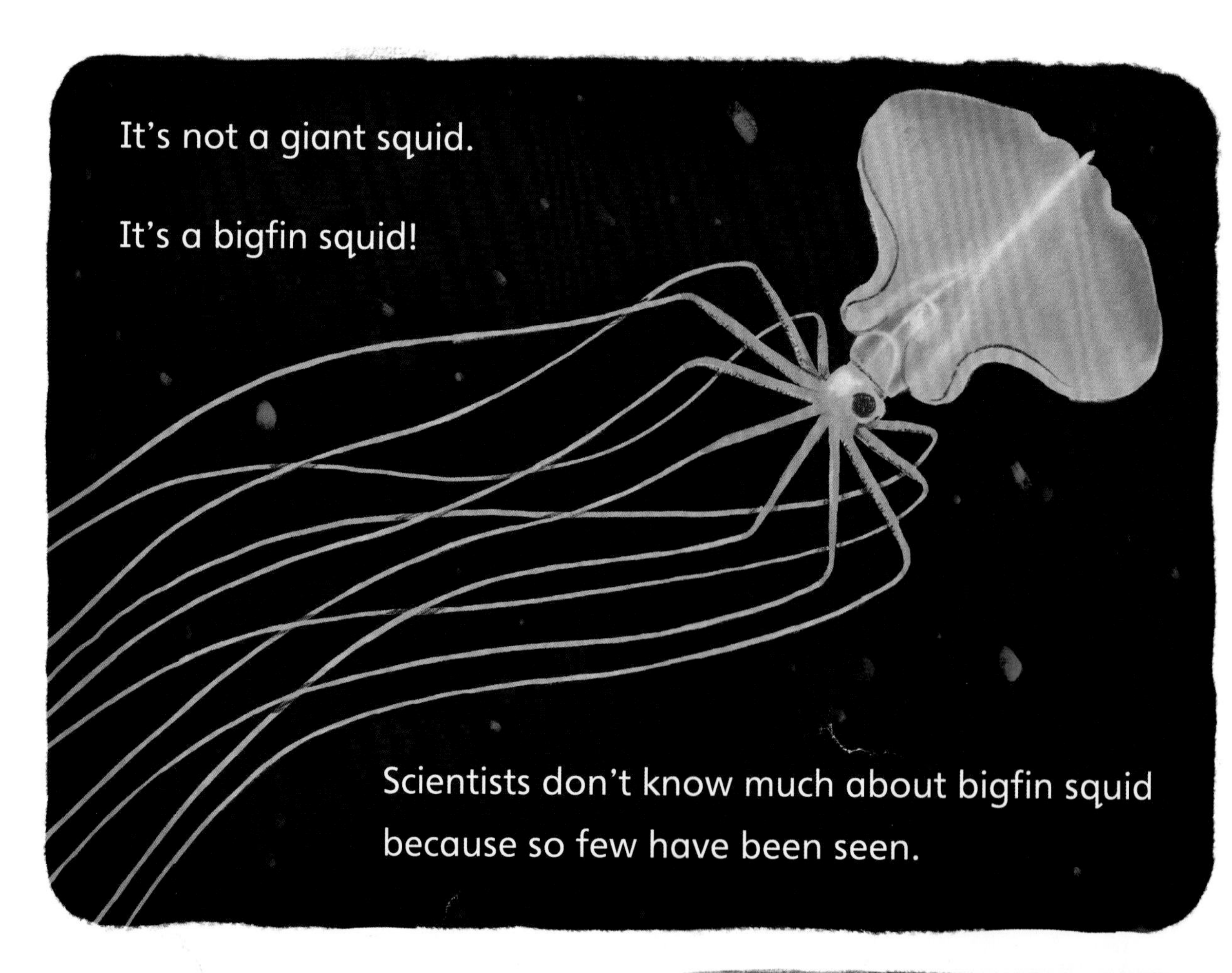

Quick, take photos and
video before it swims away!

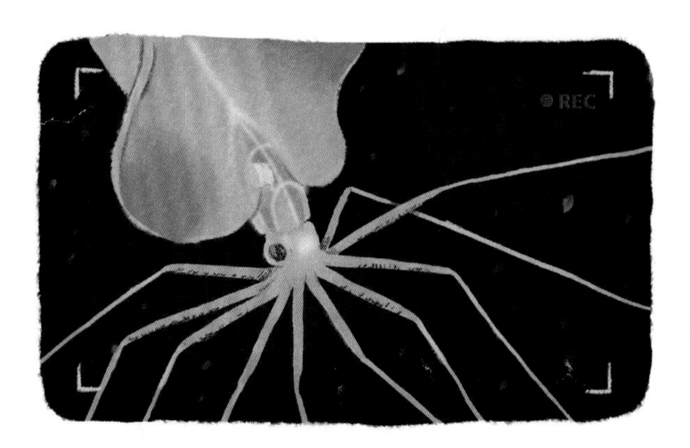

Your adventure continues on page 39.

Robin Is Your Pilot!

I can't wait to get to sea!

Do you need help picking a submersible?

A long-range sonar can help us see things farther away, like cliffs and seamounts.

Also, small submersibles are easier to navigate.

To pick a submersible, go to page 24.

The sharks swarm around *Galaxy*.

They accidentally bump your submersible as they pass.

Your pilot quickly navigates *Galaxy* away from the whale fall.

Finally, the sharks go back to eating!

Look at that!

Could it be a giant squid?

Your adventure continues on page 51.

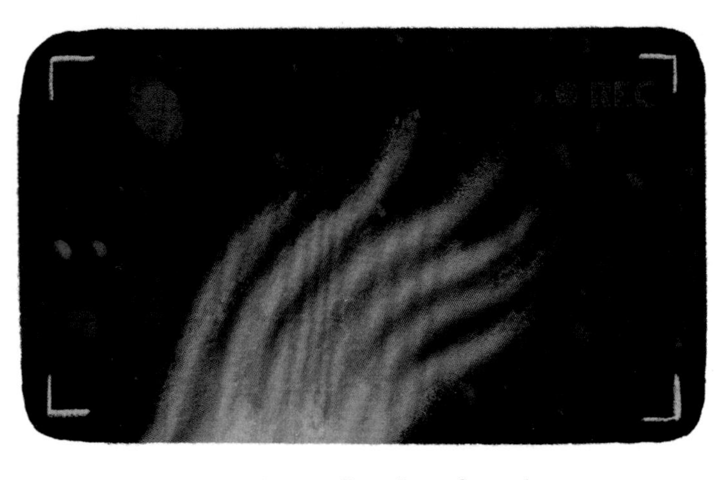

Look, a flash of red!

Your adventure continues on page 60.

It's a brine pool.

That's an underwater lake filled with very salty water.

The salt makes the water in the brine pool heavier, causing it to sink below the seawater.

You collect samples, photos, and video until you need to go back to the ship to recharge your submersible's batteries.

Many animals thrive around brine pools, but the water inside is toxic for most.

If an animal dies in a brine pool, the salty water stops its body from breaking down.

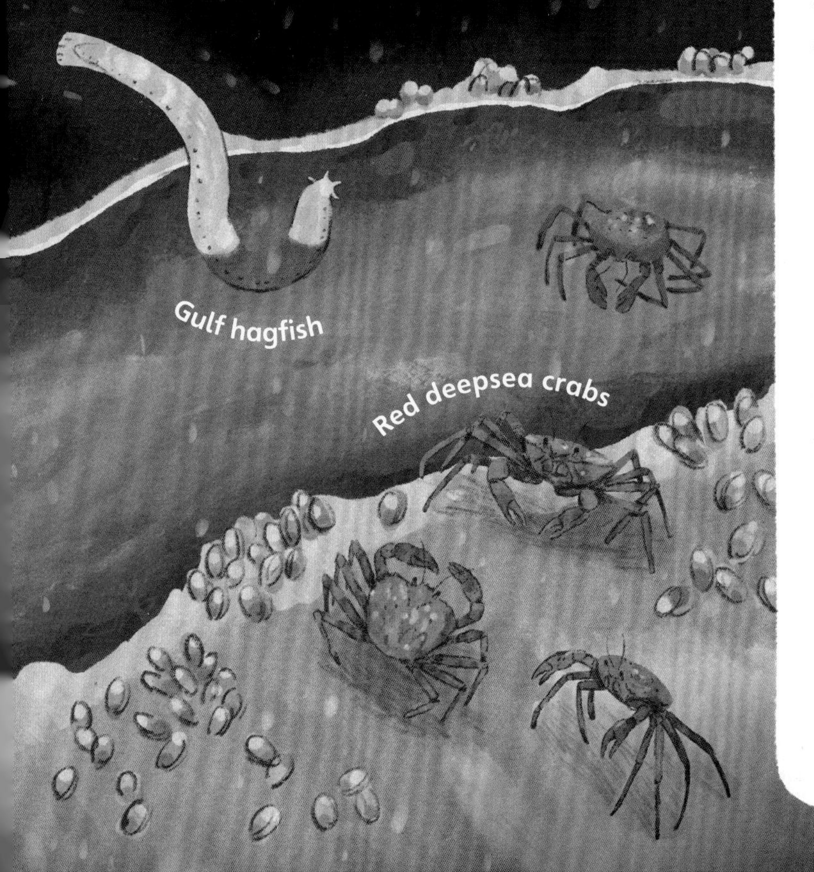

No giant squid, but you discovered this brine pool with its complex ecosystem!

End of dive.

To try again, go back to page 22.

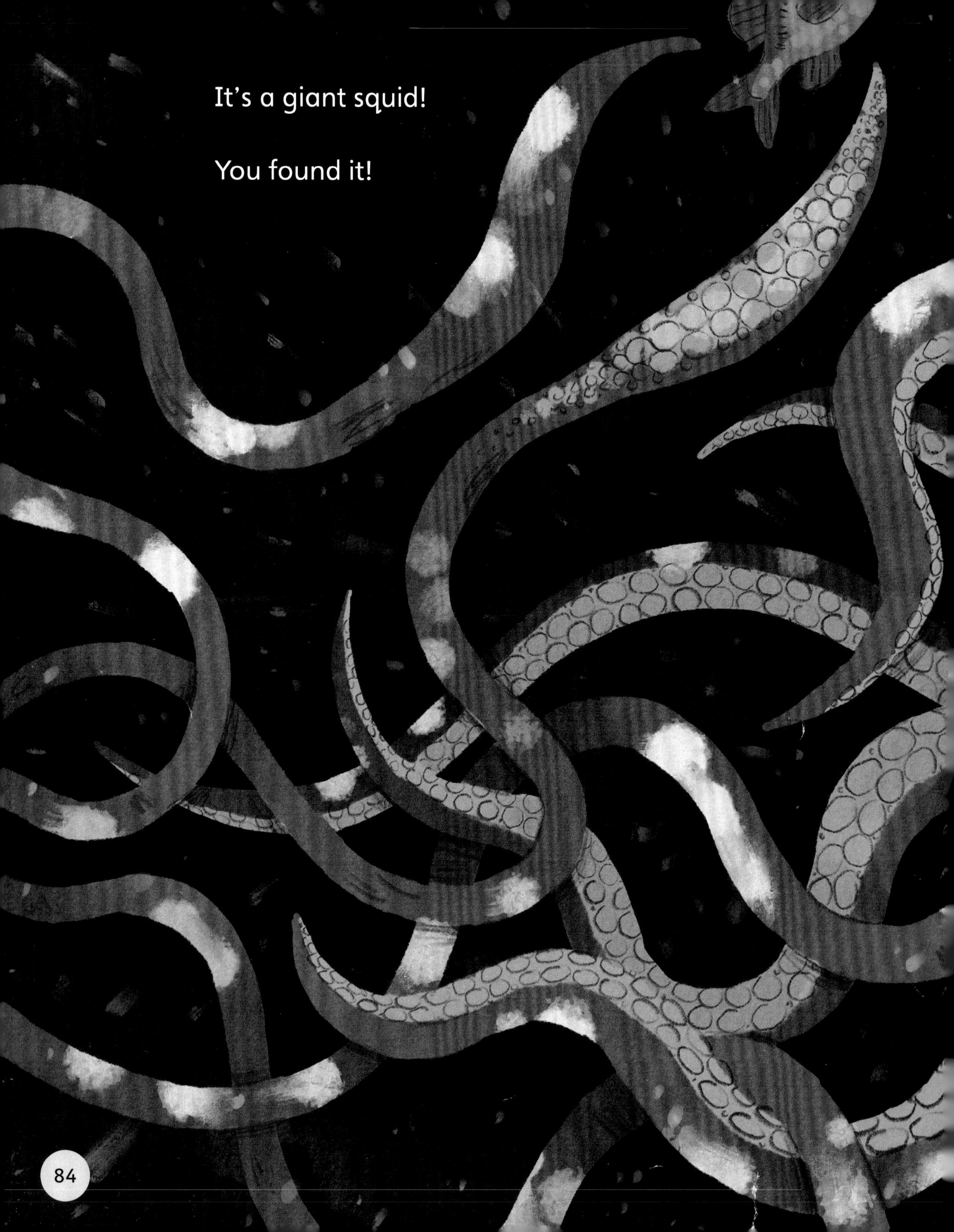

It’s a giant squid!

You found it!

Only a few people have seen live giant squid in their natural habitat.

Now you are one of them!

Congratulations!

A Note from Amy and Andy

While creating this book, we asked scientists and submersible specialists a lot of questions. We've done our best to use scientific information about squid, other ocean animals, and marine science. We've also put a lot of our imagination into this book. Any mistakes are our own.

Scientists know a lot, but there's still so much they are trying to discover about squid—giant squid in particular. How many live in the world's oceans? How do they reproduce? How are they reacting to ocean pollution?

There's still a lot to explore. Will you be the scientist who helps solve these mysteries?

Special Thanks To

Danna Staaf
Squid Scientist
and Author

Kakani Katija
Principal Engineer,
Monterey Bay Aquarium
Research Institute

Peter Girguis
Professor of Organismic
and Evolutionary Biology,
Harvard University

Rika Anderson
Assistant Professor
of Biology,
Carleton College

Sarah McAnulty, PhD
Squid Biologist,
University of Connecticut

Scott A. Hansen
ROV Pilot/Technician, ROV
Doc Ricketts, Monterey Bay
Aquarium Research Institute

Sönke Johnsen
Professor of Biology,
Duke University

Additional thanks to

Owen in Massachusetts, Chris Z. Yu, Julie Kim, Annie Carl,
Ben Clanton, Julia Kuo, Stena Troyer, Alec Chunn, Joy Zhang,
Bryan Wilson, Yuchia Kao, Andrew James Sapala, Colin Brandt,
Blake Chamness, Kimberly Hutchinson, and Metal Pig.

Animals in This Book

Search at your local library or online to find books and information about these amazing animals!

A note about names: A single species can often be called several different common names depending on who is writing or talking about them. For this book, we have relied on the names most generally used by scientists in the United States, but it is important to remember that these names have, in most cases, been selected by white European scientists rather than by Indigenous peoples who may have known about the species for thousands of years. As scientists continue to engage in conversations about colonialism, these names may change.

Common Name	Scientific Name
Atlantic pyrosome	*Pyrosoma atlanticum*
Barreleye	*Macropinna microstoma*
Benttooth bristlemouth	*Cyclothone acclinidens*
Bigfin squid	*Magnapinna* sp.
Black-eyed squid	*Gonatus onyx*
Bluefin tuna	*Thunnus thynnus*
Bluefish	*Pomatomus saltatrix*
Bluntnose sixgill shark	*Hexanchus griseus*
Brokenline lanternfish	*Lampanyctus jordani*
Brown booby	*Sula leucogaster*
Cape hake	*Merluccius capensis*
Chambered nautilus	*Nautilus pompilius*
Cold seep mussel	*Gigantidas childressi*
Deep-sea dragonfish	*Eustomias trewavasae*
Fin whale	*Balaenoptera physalus*
Giant oarfish	*Regalecus glesne*
Giant siphonophore	*Praya dubia*

Giant squid	*Architeuthis dux*
Goblin shark	*Mitsukurina owstoni*
Golden deepsea crab	*Chaceon fenneri*
Green sea turtle	*Chelonia mydas*
Grey cutthroat eel	*Synaphobranchus affinis*
Gulf hagfish	*Eptatretus springeri*
Helmet jellyfish	*Periphylla periphylla*
Hidden bristlemouth	*Cyclothone obscura*
Krill	*Euphausia americana*
Longtooth anglemouth	*Sigmops elongatus*
Manylight fangfish	*Chauliodus sloani*
Moon jelly	*Aurelia aurita*
Oceanic manta ray	*Mobula birostris*
Orca	*Orcinus orca*
Piglet squid	*Helicocranchia pfefferi*
Pink shrimp	*Penaeus duorarum*
Razorback scabbardfish	*Assurger anzac*
Red deepsea crab	*Chaceon quinquedens*
Red paper lantern jellyfish	*Pandea rubra*
Sailfish	*Istiophorus platypterus*
Sperm whale	*Physeter macrocephalus*
Spikehead dreamer	*Bertella idiomorpha*
Stout blacksmelt (or owlfish)	*Pseudobathylagus milleri*
Streaked shearwater	*Calonectris leucomelas*
Striped dolphin	*Stenella coeruleoalba*
Sunfish	*Mola mola*
Swordfish	*Xiphias gladius*
Tan bristlemouth	*Cyclothone pallida*
Thickspine roughy	*Hoplostethus robustispinus*
Vampire squid	*Vampyroteuthis infernalis*
White coral	*Madrepora oculata*
Wonderful firefly squid	*Lampadioteuthis megaleia*
Yellow-nosed albatross	*Thalassarche chlororhynchos*

Glossary

Bioluminescence: Light created by the body of a living animal, plant, fungus, or microbe

Brooding: Keeping eggs or babies safe

Cephalopod: A class or group of animals that includes squids, octopuses, nautiluses, and cuttlefishes

Ecosystem: A community of plants, animals, fungi, and microbes that work together within their environment

Expedition: A trip taken by a group of scientists to study the natural world

Microplastics: Tiny bits of plastic that can be found in all our oceans and are toxic to animals and humans

Natural habitat: The environment where a living creature usually lives

Oxygen minimum zone (OMZ): An area of the ocean with very little oxygen

Prey: A living thing eaten or killed by another living thing

Science plan: A plan for an expedition that lists dive sites, tools, equipment, reason for the dive, etc.

Seamount: An extinct underwater volcano that does not reach the ocean's surface

Sonar: Short for *sound navigation and ranging*; mostly used underwater to help with navigation and mapmaking

Species: A group of related living creatures that have things in common and can produce offspring

Tidal current: The movement of water in the ocean created by the gravitational pull of the sun and moon

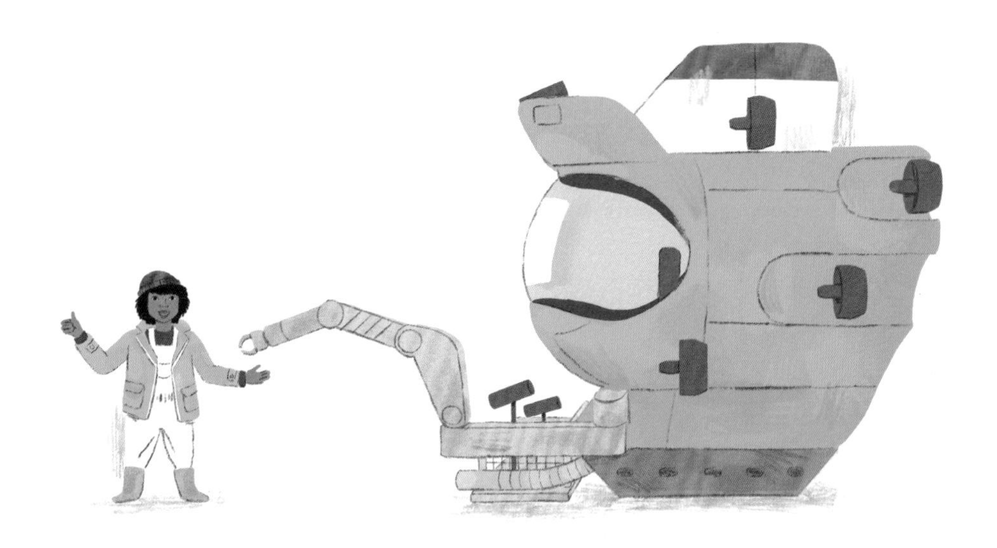

Learn More

Books

Astronaut-Aquanaut: How Space Science and Sea Science Interact by Jennifer Swanson (National Geographic Kids: 2018)

Exploring the Deep, Dark Sea (new and updated) by Gail Gibbons (Holiday House: 2019)

Flying Deep: Climb Inside Deep-Sea Submersible ALVIN by Michelle Cusolito, illustrated by Nicole Wong (Charlesbridge: 2018)

Giant Squid by Candace Fleming, illustrated by Eric Rohmann (Roaring Brook Press: 2016)

Giant Squid: Searching for a Sea Monster by Mary M. Cerullo and Clyde F. E. Roper (Capstone Press: 2012)

Glow: Animals with Their Own Night-Lights by W. H. Beck (HMH Books for Young Readers: 2015)

The Ocean Book: Explore the Hidden Depths of Our Blue Planet by Derek Harvey, illustrated by Daniel Lion, in consultation with Dr. Helene Burningham (Lonely Planet: 2020)

Websites

Monterey Bay Aquarium Research Institute YouTube. https://www.youtube.com/user/MBARIvideo.

Ocean Exploration Career Profiles, National Oceanic and Atmospheric Administration Ocean Exploration. https://oceanexplorer.noaa.gov/edu/oceanage/welcome.html.

Bibliography

Articles

Johnsen, Sönke, and Edie Widder. "Here Be Monsters: We Filmed a Giant Squid in America's Backyard." National Oceanic and Atmospheric Administration Ocean Exploration. June 20, 2019. https://oceanexplorer.noaa.gov/explorations/19biolum/logs/jun20/jun20.html.

Peterschmidt, Daniel. "Meet a 'Blue Planet' Sub Pilot." Science Friday. June 1, 2018. https://www.sciencefriday.com/articles/meet-a-blue-planet-sub-pilot/.

Ridgeon, Will. "First Whalefall Study in the Deep Atlantic." BBC. Accessed April 18, 2022. https://www.bbc.co.uk/programmes/articles/4VnhvdmDjtF4qt5Jvr26t4l/first-whalefall-study-in-the-deep-atlantic.

Roper, Clyde, and the Ocean Portal Team. "Giant Squid." Smithsonian. April 2018. https://ocean.si.edu/ocean-life/invertebrates/giant-squid.

Staaf, Danna. "Do Squid Go 'Bloop'?" Science 2.0. February 13, 2012. https://www.science20.com/squid_day/blog/do_squid_go_bloop-86905.

United Nations Environment Programme. "Microplastics: Trouble in the Food Chain." 2016. https://wesr.unep.org/media/docs/early_warning/microplastics.pdf.

Valentine, Katie. "NOAA-Funded Expedition Captures Rare Footage of Giant Squid in the Gulf of Mexico." National Oceanic and Atmospheric Administration Research News. June 21, 2019. https://research.noaa.gov/article/ArtMID/587/ArticleID/2468/NOAA-Funded-Expedition-Captures-Rare-Footage-of-Giant-Squid-in-the-Gulf-of-Mexico.

Young, Lauren J. "Untangling the Long-Armed Mystery of the Bigfin Squid." Science Friday. June 24, 2021. https://www.sciencefriday.com/articles/bigfin-squid-mystery/.

Books

Cerullo, Mary M., and Clyde F. E. Roper. *Giant Squid: Searching for a Sea Monster*. North Mankato, MN: Capstone Press, 2012.

Cusolito, Michelle. *Flying Deep: Climb Inside Deep-Sea Submersible* ALVIN. Watertown, MA: Charlesbridge, 2018.

Dipper, Frances. *The Marine World: A Natural History of Ocean Life*. Plymouth, UK: Wild Nature Press, 2016.

Hoyt, Erich. *Creatures of the Deep: In Search of the Sea's "Monsters" and the World They Live In*. Ontario, Canada: Firefly Books, 2001.

Markle, Sandra. *Outside and Inside: Giant Squid*. New York: Walker, 2003.

Nouvian, Claire. *The Deep: The Extraordinary Creatures of the Abyss*. Chicago: University of Chicago Press, 2007.

Staaf, Danna. *Monarchs of the Sea: The Extraordinary 500-Million-Year History of Cephalopods*. Lebanon, NH: ForeEdge, 2017.

Websites

Animal Diversity Web. University of Michigan Museum of Zoology. Accessed April 18, 2022. https://animaldiversity.org/.

Australian Museum. Accessed April 18, 2022. https://australian.museum/.

Blue Planet. BBC Earth. Accessed April 18, 2022. https://www.bbcearth.com/shows/blue-planet.

Blue World Institute of Marine Research and Conservation. Accessed April 18, 2022. https://www.blue-world.org/about-us/.

eBird. Cornell Lab of Ornithology. Accessed April 18, 2022. https://ebird.org/home.

FishBase. Edited by R. Froese and D. Pauly. Updated February 2020. https://fishbase.se.

Fishes of Australia. Museums Victoria. Accessed April 18, 2022. https://fishesofaustralia.net.au.

Food and Agriculture Organization of the United Nations. Accessed April 18, 2022. https://www.fao.org/home/en/.

Giant Squid. National Geographic. Accessed May 29, 2022. https://www.nationalgeographic.com/animals/invertebrates/facts/giant-squid.

Global Biodiversity Information Facility. Accessed April 18, 2022. https://www.gbif.org/.

Iziko Museums of South Africa. Accessed April 18, 2022. https://www.iziko.org.za/.

Monterey Bay Aquarium Research Institute. Accessed April 18, 2022. https://www.mbari.org/.

Monterey Bay Aquarium Research Institute Deep-Sea Guide. Accessed April 18, 2022. http://dsg.mbari.org/dsg/home.

National Oceanic and Atmospheric Administration Ocean Exploration. Accessed April 18, 2022. https://oceanexplorer.noaa.gov/.

Nautilus Live. Ocean Exploration Trust. Accessed April 18, 2022. https://nautiluslive.org/.

Oceana. Accessed April 18, 2022. https://oceana.org/.

Ocean Portal. National Museum of Natural History, Smithsonian Institution. Accessed April 18, 2022. https://ocean.si.edu/.

Photo Ark. National Geographic. Accessed April 18, 2022. https://www.nationalgeographic.org/projects/photo-ark/.

Woods Hole Oceanographic Institution. Accessed April 18, 2022. https://www.whoi.edu/.

World Register of Marine Species. Updated April 19, 2022. https://www.marinespecies.org/.

Sperm whales are
the main predators
of giant squid.

They have been found with
hundreds of giant squid beaks
in their stomachs!

A group of giant squid
is called a *school*.

Giant squid are just one
of the nearly 400 species of squids.